Water
218

海里的大象？
Elephants of the Sea?

Gunter Pauli

[比] 冈特·鲍利 著

[哥伦] 凯瑟琳娜·巴赫 绘

田　烁 译

上海远东出版社

丛书编委会

主　任：贾　峰

副主任：何家振　闫世东　郑立明

委　员：李原原　祝真旭　牛玲娟　梁雅丽　任泽林

　　　　王　岢　陈　卫　郑循如　吴建民　彭　勇

　　　　王梦雨　戴　虹　靳增江　孟　蝶　崔晓晓

特别感谢以下热心人士对童书工作的支持：

匡志强　方　芳　宋小华　解　东　厉　云　李　婧

刘　丹　熊彩虹　罗淑怡　旷　婉　杨　荣　刘学振

何圣霖　王必斗　潘林平　熊志强　廖清州　谭燕宁

王　征　白　纯　张林霞　寿颖慧　罗　佳　傅　俊

胡海朋　白永喆　韦小宏　李　杰　欧　亮

目录

Contents

一丛海藻旁边，一只墨鱼正在游来游去，这时，它突然撞见了一只海马。

　　"海马！你迷路了吗？这里可是英国海岸，你在这儿干什么呀？我记得你是生活在热带水域，而不是英吉利海峡哦！"

A cuttlefish is swimming around in a small patch of kelp when it suddenly comes face to face with a seahorse.
"Seahorse! Are you lost? What are you doing here along the British Coast? I thought you lived in tropical waters, not the English Channel."

一只墨鱼突然撞见了一只海马。

A cuttlefish comes face to face with a seahorse.

深爱着这片海藻绿洲……

"噢，其实我一直就是在这里幸福地生活呀。你看，这儿的海草'牧场'和海藻森林毗连一体，我在这里抚育小宝宝可是很安全呢。"

　　"对我们墨鱼家族来说，海藻就像是一座花园，难道你们海马家族也深爱着这片绿洲吗？"

"Oh, I happily live right here, where the seagrass fields adjoin the kelp forest, and where I can safely raise my little ones."
"So you too love this oasis of kelp that is like a garden to all of us?"

"是呀，在海藻森林里小虾随处可见，那可是我最爱的美食！还有啊，这里总是漂浮着成千上万的小块儿食物，我的宝宝们轻而易举地就能填饱肚子。他们可是这片水域里担当清洁工作的一把好手，无人能比。"

"我佩服的是……"墨鱼说道，"海藻森林那种使事物平静的力量，它能化解海浪的威力。更不用说，它还能使沉积物稳固，并释放大量氧气呢。"

"Yes, our kelp forest is full of tiny shrimp, my favourite! And my little ones can easily fill their tummies with the thousands of little pieces of food floating around here. They do the best clean-up job in the area."

"What I admire," Cuttlefish says, "is the calming power of a kelp forest, breaking the power of the waves. Not to mention that it stabilises the sediments, and pumps out a lot of oxygen."

小虾随处可见，那可是我最爱的美食！

Full of tiny shrimp, my favourite!

一条能抓住海藻枝杈的尾巴……

A tail that can grasp onto branches ...

"幸运的是，我有一条神奇的尾巴，当我在海藻之间来回穿梭时，它能把我牢牢地抓挂在海藻的枝杈上。"

"你就像只猴子！但人们却把你叫作'马'。你到底是什么呀？"

"实际上，我是一条鱼。"

"Fortunately, I have a tail that can grasp onto branches as I swing from one kelp plant to another."

"You are like a monkey! Yet they call you a horse. What are you?"

"In reality, I am a fish."

"你怎么会是鱼呢？你会通过改变颜色来伪装自己，就像变色龙那样。我甚至看到过你变成了深红色！"

"这话可不该由你来说！你叫'墨鱼'，但是你根本就不属于鱼类啊！"

"How could you be a fish? When you can change your colour too, for camouflage, like a chameleon does? I have even seen you turn a deep red!"
"Now look who is talking! You are called a cuttlefish – when you are no fish at all!"

你会改变颜色……

You can change your colour ...

我们是软体动物，不是鱼类……

We are molluscs and not fish ...

"这我知道，我来自另外一个大家族，我们家族还有鱿鱼和章鱼。没错，我们是软体动物，不是鱼类。但至少和你们比起来，我们的行为更像鱼吧！"

"噢，我不在意你们怎么称呼我们。我们海马家族有一个显著的优势，那就是我们生活在大西洋最具生物多样性的海域中。"

"没错，海藻森林确实是产卵和抚育后代的理想之地。它是迄今为止任何父母所能想象到的最佳孕育地带。"

"I know that. I am from a family that includes squid and octopus. Yes, we are molluscs and not fish. But at least we behave much more like fish than you do!"

"Oh, I don't care what you call us. We seahorses have the great privilege of living in one of the most biodiverse areas of the Atlantic."

"Yes, this kelp forest is ideal for laying eggs and raising one's young. It is by far the best nursery any parent could imagine."

"还有啊，这里的水质非常纯净，在这儿穿梭真是件令人愉快的事情。还能躲避捕食者的猎杀，安全极了。"

"但是，还有件事儿想请你给我讲讲。"墨鱼问道，"我总是对你的大鼻子感到特别好奇……"

"别不好意思嘛，继续问，你想知道什么？"

"And the water is so pure, it is a delight to swim around. And it is safe from predators."

"But please tell me something," Cuttlefish asks. "I have always wondered about your snout…"

"Don't be shy. Go ahead and ask. What is it you want to know?"

我总是对你的大鼻子感到特别好奇……

I have always wondered about your snout ...

你的大鼻子工作起来就像个吸尘器……

Your snout works like a vacuum cleaner ...

"希望你别介意，这问起来可能有些难为情……是这样的，听说你的大鼻子工作起来就像个吸尘器，这是不是真的呀？"

"绝对是真的，而且还是一款非常好的吸尘器！在其他任何人都钻不进的犄角旮旯和缝隙之中，我能探测到藏在里面的食物，只要能发现它们，我就能把它们吸出来。"

"如果有块食物比你的大鼻子还要大可怎么办？"

"If you do not mind, it might be a bit embarrassing... but is it true that your snout works like a vacuum cleaner?"

"Absolutely, and a very good one at that! I can probe for food in nooks and crannies where no one else can reach. And suck up whatever I find."

"And what if the piece of food is bigger than your snout?"

"那样的话，我的大鼻子也会变大，大到平时的两倍吧。看上去就像大象的鼻子。"

"这么说，你们海马既不是马，也不是鱼，更不是猴子或变色龙，而是在祖上和大象家族颇有渊源？也许，你们是生活在海洋里的大象吧？"

……这仅仅是开始！……

"Then my snout grows, to almost twice its size. And looks like a trunk."

"So you seahorses are not horses, or fish, or monkeys, or chameleons, but have ancestors that are family of elephants? Are you perhaps the elephants of the sea?"

... AND IT HAS ONLY JUST BEGUN!...

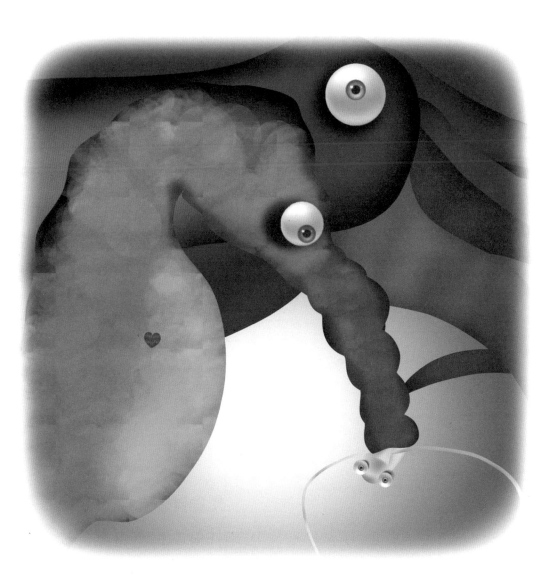

...AND IT HAS ONLY JUST BEGUN! ...

墨鱼（学名乌贼）的种类多达120多个，是无脊椎动物中最聪明的一类。墨鱼可以很好地控制自身浮力。它们中的绝大多数都生活在海洋底部，只在有限的区域内觅食和寻找伴侣。

There are 120 varieties of cuttlefish, the most intelligent of the invertebrates. Cuttlefish are able to control their buoyancy. Most live on the ocean floor, searching for food and mates in only a limited range.

墨鱼骨是一种多孔贝壳，可以用作笼中饲养鸟类的饮食补充物，我们通常可以在海滩上看到墨鱼骨。墨鱼通过调节骨骼前后腔体中的气体重量，来掌控游动节奏以及停靠的位置。

The cuttlebone often found on the beach, and used for caged birds, is a porous shell. A cuttlefish adjusts the gas in its forward and rear chamber of the bone to regulate and control where it will rest.

墨鱼游得很慢，当它们还很小时，它们更倾向于用腕足"走路"。墨鱼虽然是色盲，但仍然可以根据环境需要变成任意颜色来伪装自己。

Cuttlefish are slow swimmers, and when still small they prefer to 'walk' on their arms. Cuttlefish are colour-blind, and yet they can turn any colour required to camouflage themselves.

墨鱼可以改变身形和皮肤质地，这样就可以创造完美的伪装。它们利用自己身体上的小凸状物改变皮肤质地，重新排列自己的腕足位置以匹配周围环境。

Cuttlefish replicate shape and texture to create the perfect camouflage. They use tiny bumps on their bodies to change the texture of their skin. They rearrange their arms to match the shape of the background.

Cuttlefish have W-shaped pupils, giving them wide horizontal vision, and are able to see nearly 360 degrees around them. They also produce decoy cuttlefish to distract and confuse attackers, while they shoot away.

墨鱼的瞳孔呈 W 型，视野宽阔，能看到近 360 度的景象。它们还能产生墨鱼假体，当假体发射后，就能起到分散攻击者注意力或迷惑攻击者的作用。

The UK's ocean temperature is projected to rise by between 1.2° C and 3.2° C by 2100. The North Sea is experiencing immigration by exotic species, like red mullet, anchovy and pilchards.

预计到 2100 年，英国的海洋温度将上升 1.2℃至 3.2℃。北海正面临外来物种迁入，如红鲻鱼、凤尾鱼和沙丁鱼。

Plastic and nanoparticle pollution in the North Sea, both from fishing gear and the microscopic particles, resulting from physical and chemical breakdown, is endangering the life of many aquatic species.

北海的塑料和微粒污染，不论是来自渔具，还是经物理和化学反应后分解成的微塑料，都正在危及许多水生物种的生命。

Sewage treatment plants of current design are not capable of eliminating significant inputs of microplastic pollution into freshwater and marine water bodies.

现在的污水处理厂还无法防止大量微塑料污染淡水和海洋水体。

Can you imagine your nose doubling in size?

你能想象你的鼻子变大 2 倍吗?

Why do we so easily name animals incorrectly?

为什么我们给动物起名字时总容易犯错?

Would you be in favour of planting more forests in the sea?

你支持在海洋里建设更多的森林吗?

Where would you like to raise your kids?

你希望在哪里抚养你的孩子?

Do It Yourself!

自己动手!

Time to undertake some research. Where are the biggest kelp forests found in the world? What happened to kelp forests? What are the reasons that seaweed forests have been decimated, perhaps even to a larger degree than the tropical rain forests? We need to identify the causes of disasters like these, and then determine what can be done. Establish an inventory of all the positive outcomes of regenerating kelp forests. Now share your insights with everyone, and inspire them to take action.

来做些研究吧。世界上最大的海藻森林在哪里？海藻森林怎么了？海藻森林遭到严重破坏，也许破坏程度甚至超过了热带雨林，原因是什么？我们需要找出造成这些灾难的原因，然后想想我们能做些什么。重建海藻森林会带来哪些好的影响？列个清单吧。然后你可以和大家分享你的见解，并鼓励人们付诸行动。

学科知识
Academic Knowledge

生物学	英国的海藻森林储存了大量的碳，在能量和营养的循环中起着关键作用；钙化藻类、内生藻类和珊瑚藻类；海草有根、茎、叶，能够开花结籽；海草形成了茂密的水下草甸，在太空中都可以看到；海草为无脊椎动物、鱼类、蟹类、海龟、海洋哺乳动物和鸟类提供庇护所和食物；海马幼体死亡率为99%。
化　学	海洋酸化，由于其物理化学性质，会影响碳酸盐-碳酸氢盐平衡，主要影响钙化生物；海藻森林会造成碘排放。
物　理	重力缺失会影响海藻的生长；海面温度的变化会改变海藻森林；海水的浮力比淡水大；吸力来自扇叶旋转形成的负压。
工程学	由于海平面上升，海岸带保护势在必行，要利用海洋动力学，如沙堤和人工岛的建设；真空吸尘器的工作原理。
经济学	通过计算生态系统服务对当地经济的贡献，包括海岸保护，英国的海洋生物多样性价值为2.67万亿欧元。
伦理学	使用不正确的名字（墨鱼不是鱼，海马不是马）；因身体特征而嘲笑某人。
历　史	45万年前，英国和法国的陆地才被分开；在希腊神话和罗马神话中，海马的上半身是马，下半身是鱼。
地　理	英国有2万千米的海岸线，71万平方千米的海域，最深处达2 000米；北大西洋洋流是大西洋经向翻转环流的一部分，形成并阻止了冰河时代；特大洪水导致英国与法国陆地的分离。
数　学	用基于过程的数学模型来研究柔性水生植被，特别是海藻，对沙丘侵蚀的影响。
生活方式	嘲笑一些具有不寻常身体特征的人，这种行为在漫画中更为突出。
社会学	开放的对话可以带来有创造力的思维模式，以新颖的想法在群体中营造自信和信任的氛围；海马又被叫作hippocampus，hippos是马的意思，kampos是海怪的意思；在西方医学史上，海马被认为有助于催生母乳，治疗脱发、狂犬病和麻风病；中医用海马来治疗阳痿、哮喘、骨折和心脏、肾脏、皮肤、甲状腺等器官的疾病。
心理学	自我意识和自信是情商的基石。
系统论	在英国水域，鳕鱼和沙鳗种群遭到过度捕捞，这对海鸟种群造成严重影响；海水中的银纳米颗粒，对蓝藻和真核浮游植物的生长有严重的抑制作用，并使细菌群落组成发生显著变化；入侵物种会导致遗传缺陷，并将寄生虫和病原体传染给本地物种；海藻森林是海洋生物的育儿所，还能够保护海岸；海底生态系统中的滤食动物形成了珊瑚礁，进而为更多其他生物创造了家园，也为幼鱼提供了良好的觅食场所。

情感智慧
Emotional Intelligence

墨鱼

海马应该只生活在热带地区，所以当墨鱼看到海马出现在英国海岸时，惊讶之情溢于言表。他直接地观察海马，并和海马探讨海马应该属于哪类动物。他关注海马的行为，指出海马和许多其他动物有相似之处。当受到质疑时，他为自己辩护说，墨鱼比海马更像鱼。墨鱼想知道海马的大鼻子是怎么回事，但又不好意思开口去问。当海马鼓励他提问时，他感到很惊讶，海马竟然没有被这个问题冒犯。她甚至还赞同了墨鱼的观察，她的大鼻子就像一个吸尘器。这又鼓励墨鱼提出了另外一个问题，试图弄清楚海马到底是什么动物。

海马

海马不会因为墨鱼的评论、批评或提出的隐私问题而生气。她关注生活在海藻森林的积极方面。面对冒犯性的评论时，她会直接反问墨鱼，为什么墨鱼不是鱼而被命名为鱼。她认为生活在一个拥有生物多样性和纯净水质的地方是一种幸运，在这里她能很好地抵御捕食者。当墨鱼的问题变得非常涉及隐私时，她鼓励墨鱼问更多的问题来满足墨鱼的好奇心。海马是完全开放的状态，没有防备，她向墨鱼提供详细的解释，以帮助墨鱼更好地了解海马所属的物种。

艺术
The Arts

通过海马，我们发现了有一种动物是真正的"变形金刚"。它既不是马，也不是鱼、变色龙、猴子或大象，但它的鼻子可以像吸尘器一样运转！发挥你的想象，画一个动物吧，让它具有上述动物的特征。你可能会画成一个怪物，但从另一方面看，这或许会是一种非常有趣的表现形式。

思维拓展
Systems: Making the Connections

我们有很多预先设定好的想法，把每个人、每样东西都放在一个特定的框架里。墨鱼不知道海马能在英国沿海水域繁衍生息，所以当发现它们的存在时感到惊讶。很多时候，当我们面对现实生活中的新信息时，我们的第一反应是嘲笑那些分享新事实的人，或者拒绝接受这些事实。在面对新信息时，我们与其感觉被冒犯，还不如将其视为一个机会，开始审视周围不断变化的现实。我们应该抓住每一个机会，通过洞察我们丰富多彩的生活和周围生物多样性的宝库，来拓展我们的知识储备。海马不仅看起来奇怪，而且行为也很奇怪。它与马几乎没有共同之处，却被人们命名为马。人们常常根据动物的单一特征，或根据其一般的外表或行为，错误地命名动物。我们应该给别人探索的自由，而不是感觉被冒犯。那些想学习的人应该得到鼓励，而不是嘲笑。我们应该在对话中提问和互动，进而了解不同的观点，还是仅仅停留在自己肤浅的观察中，不再探索更多奥秘？墨鱼的故事为我们提供了一个真实的写照，告诉我们对生物多样性的探索如何使我们认识并歌颂每一个生命在生态系统中的位置和作用。只有当我们放下固有的观念和先入为主的想法，我们才能发现更多，学到更多。我们可以在自己所处的现实环境中开始这样做。人们期望看到热带雨林和海洋中丰富的生物多样性，但我们应该花时间去寻找我们周围物种的美丽和多样性，并尝试去识别它们在生物圈中独一无二的特征、位置和作用。

动手能力
Capacity to Implement

你家或学校附近有没有什么地方可以让你更多地了解生物多样性？让我们探索一下我们周围的地区，看看我们能找到什么简单而又令人惊讶的东西。寻找鸟类、鱼类、爬行动物、藻类、真菌、蜜蜂和蝴蝶。这些东西有什么看上去奇怪的特征吗？专注观察一下，找出这些特征的作用。写一份报告来记录每个物种都教会我们什么。与他人分享你的发现吧，鼓励他们采取行动，保护他们眼前的环境——不仅是为了现在，而且是为了子孙后代。

故事灵感来自
This Fable Is Inspired by

萨拉·瓦尔德
Sarah Ward

　　萨拉·瓦尔德于 2011 年获得海洋生物学和生物海洋学理学学士学位，2015 年获得班戈大学（英国）海洋环境保护硕士学位。她进行了多项水下调查，通过研究塞舌尔、开曼群岛和柬埔寨周围的鱼类和珊瑚来评估海洋环境的健康状况。回到英国后，她受雇于位于西萨塞克斯郡的萨塞克斯野生动物信托基金会。对于英国海岸和周边的海洋野生物种多样性，她提出了许多令人惊叹的见解。

图书在版编目(CIP)数据

冈特生态童书.第七辑:全36册:汉英对照 /
(比)冈特·鲍利著;(哥伦)凯瑟琳娜·巴赫绘;
何家振等译.—上海:上海远东出版社,2020
ISBN 978-7-5476-1671-0

Ⅰ.①冈… Ⅱ.①冈…②凯…③何… Ⅲ.①生态
环境–环境保护–儿童读物—汉英 Ⅳ.①X171.1-49

中国版本图书馆CIP数据核字(2020)第236911号

策　　划　张　蓉
责任编辑　祁东城
封面设计　魏　来李　廉

冈特生态童书

海里的大象?

[比]冈特·鲍利　著
[哥伦]凯瑟琳娜·巴赫　绘

田　烁　译

记得要和身边的小朋友分享环保知识哦!
八喜冰淇淋祝你成为环保小使者!